THE BIGFOOT FIELD JOURNAL

(Pacific Northwest)

J.P. Riley

CONTENTS

Copyright © 2019 J.P. Riley

ISBN-13: 978-1-7941-4646-4

TERMS AND LEGAL NOTICES

The Names, characters, places and incidents mentioned in this book, are the product of the author's memory of actual accounts and experiences to the best of his knowledge.

All of the information and opinions expressed in this book are subject to change over time as science advances.

1

TILLAMOOK FOREST

It was a cloudy Friday afternoon in northern Oregon as we made our way up the windy highway through Tillamook State Forest. The rain had let up that morning and the sky was blanketed with a thick cover of clouds. We had about two hours of sunlight left as we pulled off onto an old gravel road before making our way to our final destination. We found a nice spot to park, grabbed what we needed for a light hike and headed out on foot.

We didn't have much daylight left and wanted to get as much scouting as we could done before it got dark. It was our last day in Oregon and we had a long drive home ahead of us.

We walked for a short spell along side the gravel road before I noticed a rather large game trail that was leading into the dense timber. The trail was about two feet wide in some spots and looked as if it was used recently. I quickly noticed a few large ferns that had been stepped on and other foliage that was pushed

down. This is a definite clue that some-
one or something had came through recently. I noticed a couple sets of fresh
elk prints along the trail but the rain had
washed most of them away.

◆ ◆ ◆

The trail had indicators that it had
been used for quite sometime because of
how impacted it appeared but it was still
somewhat hidden and hard to spot. I
had a lot of experience tracking animals
through the woods so it wasn't too hard
for me to follow. We slowly made our

way up the game trail and stopped for a minute to look and listen for anything that may be watching us ahead. I looked through the trees first as far as I could see then scanned the ridgeline ever so slowly.

We didn't see anything so we decided to continue on. The trail led us through a small field of ferns that led us into a dense forested canopy. I carefully scanned through the trees again before slowly making our way into the forest.

The ferns were very green and some of the trees were covered with a heavy layer of bright green moss. It was a beautiful forest that had a somewhat majestic feel to it. The forest was very colorful and different than any other forest that I've ever seen.

The rain had softened up the ground which made for some nice quiet walking. We slipped into the forest and made our way in about ten yards to a level area where we could scan our surroundings. As I looked through the trees I could hear the faint trickle of a small stream that was running down the ridge. The trees were widely spaced out which made it easy to see in all directions.

The forest appeared to be very thick from the road but once we got inside it sort of opened up into a majestic looking rain forest like I have never seen before.

2

SCOUTING FOR SIGN

onestly I wasn't really planning on seeing or hearing any sort of Bigfoot activity or sign. I had never been to this place in Oregon before and

I have always been quite a skeptic when it comes to Bigfoot living in areas that weren't that remote. At that time I really didn't know anything about the Tillamook State Forest. I never heard of any Bigfoot sightings occurring there either until that week. Despite how I felt I decided to continue scouting around the area. I canvassed the ground as we slowly worked our way further up the ridge for any and all tracks that I could spot. It had been raining a lot so most of the older animal tracks were washed away. I saw a marsh about 20 yards up

the hill and slowly made our way there.

W e noticed a few larger sized tracks that were bigger than any deer or elk could have made. They looked like they were in the shape of a human foot but there really wasn't any-thing that stood out to me as a positive Bigfoot track! As we made our way to the marsh I noticed a few larger prints that were sporadically scattered around the area. I stopped and investigated each one before moving on to the next. When we got to the marsh I found a large print that was a bit larger than the size of my foot. The print somewhat older and was full of water. It was hard to classify it because of how old and wet it was. It must have rained a lot because there wasn't much detail left in the print so we moved on. I looked around, scanning the plants and trees in the forest around us as very carefully. Every little tle fern that was broke and pushed over told a story as to what could have possibly came through there and reacted to the environment. Despite going into the area not really thinking that I would see any sign of a Sasquatch I still tried to be as thorough as I could. All animals including primates leave a mark on the environment everywhere they go. Sometimes the most subtle clues can be an indicator of something big! I noticed an area in the small fern clearing that we passed through earlier that was all trampled and looked as if something had been sleeping there. Maybe a small

herd of elk took a nap there at one point.

There wasn't a for sure way of tell-

ing what it could have been that

pushed those ferns down but finding elk tracks there helped tell a story and for me that was enough to write that off as being elk and move on.

◆ ◆ ◆

After finding some larger sets of prints around the marsh I decided to try some vocalizations. I gave about two vocalizations while standing along the side of the marsh and listened for about five minutes. We didn't hear any sort of response so we walked a little further up the ridge.

My girlfriend was about fifteen yards behind me filming with her camera, when she signaled me to come check out something that she found. I walked back down to her and noticed a large bone that was laying on the ground. The bone had some green moss like color to it and looked like it was still somewhat fresh. Bones that have already went through their initial decay cycle usually turn all white in color after all of the fleshy like substances are usually all but gone.

The bone still had a good amount of

fleshy like substance on it which told me

that it hadn't been there too long. Maybe

a few months! If I had to guess I'd say

that it looked like a leg bone, possibly from an elk. Maybe a predator carried it there and was feeding on it. Where did that bone come from? Was it from an elk that was killed during the past hunt-ing season or could it have been killed by a predator?

I'm not a forensic scientist and I wasn't really sure what animal the bone could have came from! Could the bone be from a human or a Bigfoot? I highly doubt it but never the less, we decided to leave it there. I plan on taking the pictures of the bone to a forensic anthropology professor at a local college to see what they think. I didn't notice any other bones or hair sitting nearby which led me to believe that a predator likely carried the bone there and left it.

3

ROCKS AND WOOD KNOCKS

After we got enough pictures of the bone I found a couple hand sized rocks that caught my eye so I decided to check them out. It's known that primates use wood knocks to communicate with one another. I believe that they do this to keep track of where each one is and a lot of other things that we may never know or understand. If primates use sticks for wood knocking why not try to use stones and bone as well?

The wood knocking that I heard last summer while hiking up at a place called Wellington sounded different than any sounds that I've been able to replicate using wood alone. It was almost like a bone and wood together or maybe even stone!

If Bigfoot does exist, could it be possible that they use a combination of wood, stone and bone to wood knock with? Anything is possible so why not do everything I can to find out? I often wondered if Bigfoot choose a certain bone or stick to carve or doctor up in a unique way to produce the woodknocking sounds. If so, do they carry it with them throughout their life span or until they find the need to make another one? Is there some complex level of behavioral traits amongst bigfoot in which they make tools for wood knocking and pass them on to their offspring?

♦ ♦ ♦

I think it's very possible given how intelligent our modern day primates seem to be. Look at apes and mountain gorillas for example. They have many different intelligent primate characteristics that intrigue most scientists. When I was attending college at the University of Washington, I took a class on primates. In that class told us a story about a chimpanzee named Kanzi. He was somehow taught to communicate with researchers using flashcards and a keyboard or something like that. He was one very intelligent chimpanzee. In a case where we're dealing with a primate that's nine feet tall, walks upright like a human and has some pronounced Neanderthal traits, it would be very unscientific to assume that primates aren't smart enough to forge tools for wood knocking. I find all of this to be very fascinating!

I grabbed the two stones that I found and sort of clanged them together. At first I hit the two stones together once. Then I did two smaller clangs and repeated those steps for about ten seconds. I paused for about five minutes and did it again. I've always felt that it's important to minimize your calls as much as possible when out in the field because it can make them sound more authentic. If you've ever tried elk calling it's similar in a some ways. The elk might know something's up if you over exaggerate on your calls causing the elk to run away. Sometimes less is better. Based on my experience I found this to be mostly true. Even when using a predator call it seems to work the same way. That's just my opinion so don't quote me on it. Until we can prove without a shadow of a doubt that Bigfoot does exist, anything's possible and nothing is set in stone.

4

THE RIDGE

After trying my hand with the stone and having no luck or any sort of response we decided to venture further up the ridge. As I looked towards the upper part of the ridgeline I could see what looked like a lush oasis and to me it looked very interesting.

I had a feeling that it would be a good vantage point to view the other side of the ridge and I was anxious to get up there and check it out. We took our time making our way up the ridge in hopes of not scaring anything off. My girlfriend was about ten feet behind me filming and looking for signs as well. When I got to the top I could see down the other side and into a small draw. To my right, the ridgeline went up further into a lush green area. It felt like I was walking into a jungle! It was so beautiful and the environment felt almost apelike so I decided to name it the Ape Forest!

As I waited for my girlfriend to catch up, I found a good sized stick to use for wood knocking. I got a snug grip on the stick with my right hand then proceeded to strike a nearby tree a few times. Then I paused for a brief moment and listened intently for any sort of response.

We both stood there for about five minutes looking for any movement and listening for any kind of response. After not getting any sort of response, we decided to move on. My hopes of finding anything that could possibly indicate a Bigfoot being in the area were dwindling fast!   We noticed some cool looking fungi that was growing off the side of a tree so we took some pictures of them before we decided to start heading back.

We had been hiking for about an hour and a half and wanted to make sure we made it back before dark.

We slowly made our descent from the ridge back into the forest valley below.

After about five minutes of walking I stopped to let out another short series of vocalizations. We listened intently for about five minutes and after having no response again we decided to keep moving. As we continued to make our way down the ridge we took a short break near the marsh area where we found the large bone earlier.

I scanned the area again for any sort of possible signs of a large primate. It was in that instant that I saw something that really caught my attention. My eyes perked up when I noticed a large track on the side of the marsh that looked somewhat fresh.

It was about ten yards away from where we were standing so we quickly made our way over for a closer look. As I got closer my heart started to race. With each step I took, I could start to see what looked like the outline of toes and other small details of a print that the rain hadn't washed away!

The print itself was larger than any human could possibly make! I could see what looked like toes, a heel and a few other parts of a foot. Bears don't normally leave prints that big. I've seen many different bear prints in the wild but this was neither human or bear. Nothing in those woods could have made a foot print that large.

After we finished inspecting the print I noticed another large print about three and a half to four feet away. As we slowly stood up to inspect the other print the distinct sound of wood knock-ing sounded off from the top of the ridge! I was lucky enough to record the two wood knocks on my audio recorder as well.

Keep in mind that there weren't any houses nearby, no people hiking around and no possible sign that anyone was up on the ridge that day! As I listened to the wood knocks on my audio it sounded a little different than any noise I was able to reproduce using just a stick. It's hard to describe but it almost sounded like bone and wood together or possibly even stone!

It was so bizarre but very intriguing at the same time! What would possibly be used to make such a noise? The crazy part is that the noise sounded just like the wood knocks that I heard up at Wellington last summer! We didn't hear any more wood knocks after that nor did we see anything come down the ridge.

After we finished inspecting the first print we walked over and examined the other track as well. It was also very large in size. The ground was somewhat hard so it would have taken something with a lot of weight and size to leave such and indentation in the ground like those prints did!

We took photographs of the prints and made our way back to the gravel road before dark.

5

OVERVIEW

After we made it back to our car I was already wanting to check out the video footage of our trip. Honestly I had no idea that we were going to hear what we heard and find the kind of prints that we did at Tillamook State Forest. I went there feeling like it was going to be a big waste of time or a good hike at the most. Seeing those prints and hearing the wood knocks gave me hope that there could very well be a large primate or two thriving in that part of Oregon. That also lends a small bit of credibility to the other Bigfoot encounters that I've heard about in the Tillamook State Forest as well.

The Tillamook forest is made up of over 350,000 acres and is home to many different types of large game animals. Could there be a large unknown primate living in the remote parts of Tillamook State Forest? I think it's very possible!

◆ ◆ ◆

I decided to write a Bigfoot field journal book series so that I could start documenting and sharing my adventures with the world! This whole experience was very exciting and I hope that you enjoyed reading this book. I'd also like to hear about your encounters with Bigfoot as well. Feel free to leave me a comment. Thanks again for reading my books and thanks for all of your support. Best of luck in your Bigfoot adventures.

Thanks again and best of luck in your Bigfoot adventures!
J.P. Riley

* 9 7 8 1 7 9 4 1 4 6 4 6 4 *